CECIL COUNTY PUBLIC LIBRARY
ELKTON, MARYLAND 21921

SEP 2 4 1999

D0844244

YOUNG SCIENTISTS

Sound and Music

Published by
Smart Apple Media
123 South Broad Street
Mankato, Minnesota 56001

Cover design by Patricia Bickner Linder
Interior design by Neil Sayer

Photographs by Bruce Coleman Limited,
Image Bank, Getty Images,
Robert Harding Picture Library

Copyright © 1999 Smart Apple Media.
International copyrights reserved in all countries.
No part of this book may be reproduced in any form
without written permission from the publisher.

Printed in Hong Kong.

Library of Congress Cataloging-in-Publication Data

Dixon, Malcolm and Karen Smith.
Sound and music / by Malcolm Dixon and Karen Smith.
p. cm. — (Young scientists)

Includes index.
Summary: Explains the production of sounds and music
and provides simple experiments to illustrate the
principles described.
ISBN: 1-887068-72-4
1. Sound—Juvenile literature. 2. Sound—Experiments—
Juvenile literature. 3. Music—Acoustics and physics—
Juvenile literature. 4. Music—Acoustics and physics—
Experiments—Juvenile literature. [1. Sound. 2. Music.
3. Sound—Experiments. 4. Music—Experiments. 5.
Experiments.] I. Smith, Karen, 1958- II. Title. III. Series:
Dixon, Malcolm. Young scientists.

QC225.5.D59   1999
534—dc21   98-4218

First edition

2 4 6 8 9 7 5 3 1

YOUNG SCIENTISTS

Sound and Music

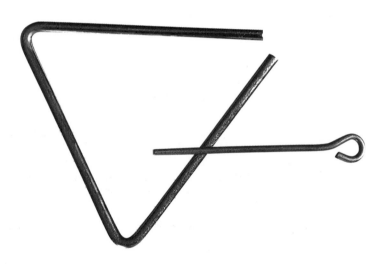

**Malcolm Dixon
and Karen Smith**

Smart Apple Media

NOTES FOR TEACHERS AND PARENTS

Sounds around you (pages 6-7)
Try to make children more aware of the varieties and sources of sounds around them.

Sound is a vibration (pages 8-9)
Encourage children to use the word "vibration." Let them use rulers and tuning forks to experience vibrations. Tuning forks can be purchased from music stores. The effects of vibrations can also be seen when rice is sprinkled on a drum. When the drum is hit, the surface vibrates, causing the rice to jump into the air.

Sounds travel (pages 10-11)
Use the photograph and the apparatus to develop the idea that sound travels in invisible waves from its source through the air (and through solids and liquids) to our ears. Remember that sound waves travel in all directions; they are three-dimensional.

Can you hear it? (pages 12-13)
Use the photograph to develop the idea that sound travels away from the source, in this case the triangle, and gets fainter as it does so. If possible, let children investigate their ideas firsthand. When using the radio as a source of sound, try to develop the idea of a "fair test," i.e. the volume of the radio should be the same for both parts of the investigation. Discuss the idea that sounds can be heard through the tube because the vibrations are prevented from spreading out. You may need a longer tube than the one in the photograph.

Hearing sounds (pages 14-15)
Remind the children never to insert anything into their ears. The ear drum is easily damaged. When using sharp needles or pins, children need to be carefully supervised. An extension activity would be to make a cardboard ear trumpet (cone-shaped) and to see if this enabled the children to hear better. Let them observe animals—a cat, for example—and discuss how they funnel sounds down their ears.

Making sounds louder (pages 16-17)
Warn children of the possibility of damaging their ears through loud sounds, especially when using headphones. Demonstrate increasing the volume (loudness) of a radio. If it is difficult to hear heartbeats through the ribcage, the children should run in place for a couple of minutes. Discuss how a doctor uses a stethoscope to listen to sounds in the body.

Your voice (pages 18-19)
Let the children feel for their larynx. It is more prominent in some individuals than in others. When they say "aaargh," they may be able to feel the vibrations in their throats.

The sound of music (pages 20-21)
Music is made when regular vibrations are made at regular intervals. The instruments referred to here are percussion instruments such as drums, cymbals, triangles, bells, gongs, tambourines, and so on. The plant pots can be suspended from the string by using small nails fixed horizontally. The larger pots produce lower notes because they vibrate more slowly.

Blowing musical notes (pages 22-23)
Let children carry out the bottle investigation for themselves. Food coloring can make the water easier to see. Can they make a short tune? Some children may play musical instruments—flute, recorder—which they can demonstrate to others. Use magazines, catalogs, and books to collect photographs of wind instruments. When making the wind instrument, moving the cork alters the length of the air column, thereby varying the notes produced when blowing.

Vibrating strings (pages 24-25)
If possible, let the children have some first-hand experience with string instruments. Discuss the vibrating strings and the sounds they make. Invite a guitarist or violinist into school to play for the children. Adult help is needed to cut the cardboard on the model guitar. Stretching the bands tighter will make higher sounds. Let the children investigate by using different thicknesses of rubber bands and by varying the lengths of the bands.

Sending sounds (pages 26-27)
When making the telephone, use "hard" string and keep it taut. The sound travels in the form of vibrations along the string to the ear. If the string is loose, it cannot produce the vibration necessary to send a sound.

Bouncing sounds (pages 28-29)
Find a safe place where the children can experience echoes for themselves, such as a large empty room, a tunnel, subway, or similar enclosure. Discuss how the sound travels as invisible waves and bounces off the walls back to their ears. When carrying out this experiment, you may use the word "reflect" as well as "bounce," i.e. the sound is reflected from the stiff cardboard.

 Contents

Sounds around you

There are many kinds of sound, and they are made
in different ways. Close your eyes. Keep still and
listen. What sounds can you hear? Can you hear
people talking? Can you hear the sounds of cars?
Can you hear children playing? Can you hear any
music? What can you hear?

The world is full of different sounds.

Work with a friend

Use a tape recorder to make a "sound quiz." Try to include some of these things on your tape:

You will need:
small tape recorder

a dog barking   a door closing
children singing and clapping footsteps
water from a faucet   a clock ticking
music from a radio

What other sounds could you include on your tape?

Now see if your friends can guess what the sounds are.

Sound is a vibration

Sounds are made by movements called vibrations. Every sound you hear is made by something that is vibrating.

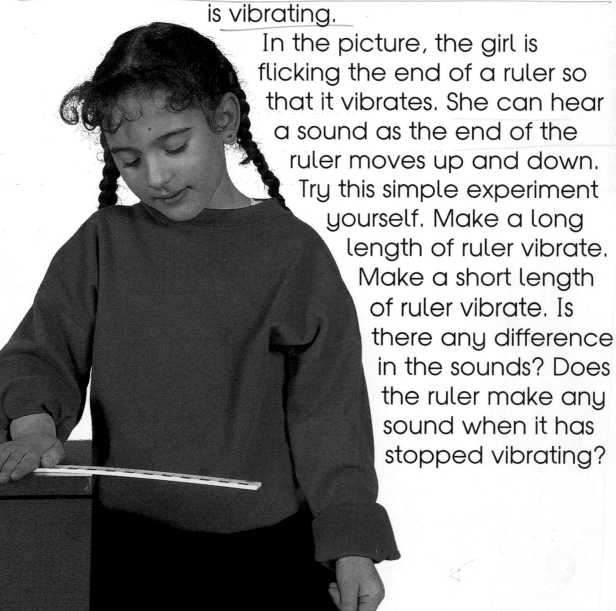

In the picture, the girl is flicking the end of a ruler so that it vibrates. She can hear a sound as the end of the ruler moves up and down. Try this simple experiment yourself. Make a long length of ruler vibrate. Make a short length of ruler vibrate. Is there any difference in the sounds? Does the ruler make any sound when it has stopped vibrating?

Feel the vibrations!

Strike a tuning fork gently on a book.
Listen to the sound it makes.
Strike the tuning fork again.
Hold it so that it gently touches your skin.
Can you feel it vibrating?

You will need:
tuning fork
ping-pong ball
string
tape

Attach the string to the ping-pong ball with some tape.
Strike a tuning fork and hold it next to the hanging ping-pong ball.
What happens? Why does this happen?

Sounds travel

Throw a stone into a pond, and you can see small waves on the surface of the water. The waves spread outward from the place where the stone hit the water.

Sounds also travel in waves. Sound waves are invisible. They push through the air from the object making the sound to your ears.

Work with a friend
Make a wave!

You will need:
2 long lengths of tape
wooden popsicle sticks

Stretch a length of tape—sticky side up—across a table top. Place popsicle sticks across the sticky surface. Now stretch another length of tape—sticky side down—on top of the sticks.

Ask a friend to hold one end of the tape while you hold the other end. Tap one of the wooden sticks with a finger. Watch how the "wave" moves across to your friend.

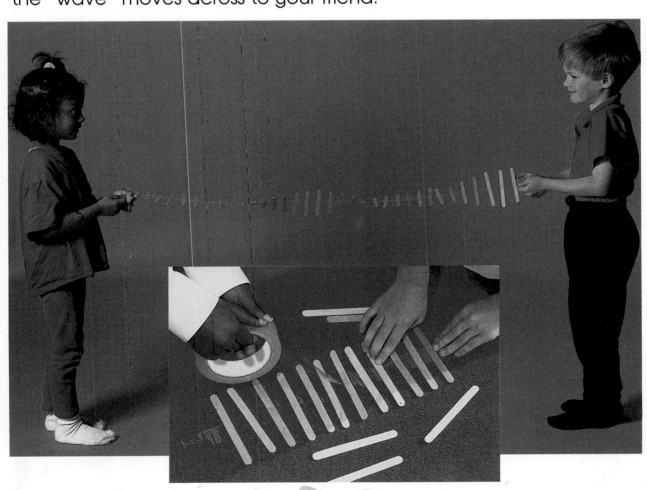

Can you hear it?

Look at the picture. The teacher is striking a triangle. Which children can hear the sound of the triangle? Who can hear the loudest sound? Is the sound softer for some of the children? Are there some places where the sound can not be heard? Test your ideas.

Work with a friend
Now try this

Turn the radio on so that it is playing quietly. Move away from the radio so that you cannot hear it play. Now walk toward the radio and stop when you can hear it. Mark this spot.

Now make a cardboard tube the same length as the distance from the radio to the spot you have marked. Listen to the radio through the long tube.

Can you hear the radio playing? Is it louder, softer, or the same as it was before?

Why does this happen?

You will need:
small radio
cardboard tubes
tape

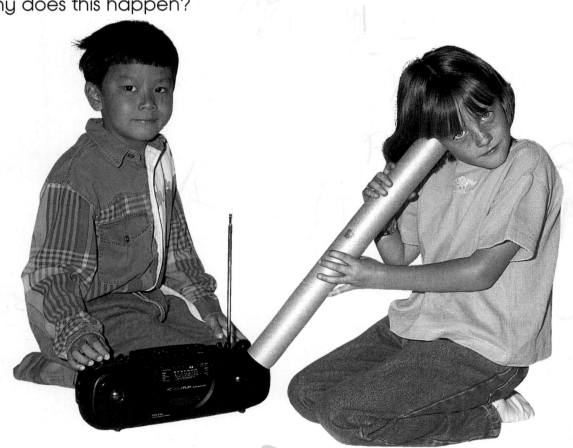

Hearing sounds

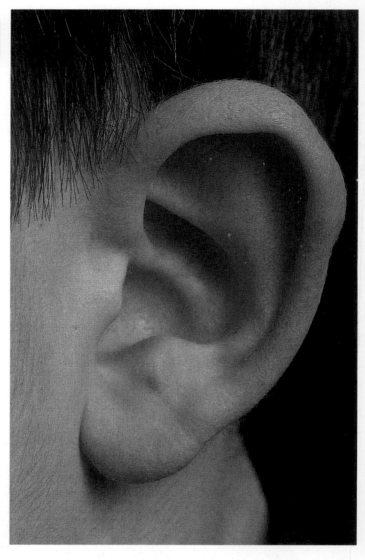

You hear sounds when they enter your ears. The part of your ear that you can see is called the outer ear. The shape of the outer ear enables it to collect sound waves. The sound waves then pass down a tube to delicate parts of your ear that are inside your head. Inside your ear, the sound waves cause the eardrum to vibrate. From your eardrum, sound messages are sent to your brain.

Work with a friend
Can you hear a pin drop?

You will need:
pin or needle
metal dish
quiet room

Ask your friend to stand a long way behind you. Then ask your friend to drop the pin onto the metal dish.

Did you hear the sound of the pin dropping? Let your friend move a pace toward you and then drop the pin. Did you hear the sound?

Your friend should move forward, one pace at a time, until you can hear the pin drop.

How many paces are there left between you and your friend?

Making sounds louder

These children are listening to music on a stereo system. They can turn a knob to make the music louder. The sound comes out through loudspeakers. Can you see the loudspeakers in this picture?

Radios and televisions also have loudspeakers in them. When you listen to someone talking on the telephone, you put your ear next to a loudspeaker.

Search for sounds

Cut out a square of cardboard.
Roll the cardboard to make a cone shape.
Use tape to hold it in shape.
Cut a hole and insert the tubing. Make a second cone and attach the tubing to it.

Hold one cone next to your ear and put the other cone next to a ticking clock. Is the ticking much louder now?

You will need:
thin cardboard
scissors
plastic or rubber
  tubing
tape
ticking clock

Find out more!
Listen to other quiet sounds such as your heart beating, someone whispering, and someone speaking on the telephone. What other quiet sounds can you make louder?

Your voice

You can make lots of different sounds with your voice. Your voice comes from a part of the throat called the larynx. Inside the larynx are flaps called vocal cords. When you speak or sing, your vocal cords vibrate. Your mouth and tongue help to make the sounds we hear.

Collect the sounds of human voices

Think of all the different kinds of sounds people can make. Here are some of them:

talking	singing
shouting	crying
sneezing	coughing
yawning	whispering

Can you think of more?
Use a tape recorder to make a collection of these different human sounds.

You will need:
small tape recorder

Find out more!
Make a tape collection of sounds made by different animals.

The sound of music

One way to make the sound of music is to sing. We also use musical instruments. The people in the picture are hitting their drums and making loud sounds. They hit their drums with sticks. The sticks make the skin of the drums and the air inside vibrate, causing the sound. Can you think of other musical instruments that make sounds when they are hit, tapped, or shaken? Make a display of some of these instruments.

Make a plant-pot xylophone!

Hang the plant pots from the wooden pole.
Look at the pictures to help you.
Tap each pot with a pencil.
What differences do you notice in the sounds they make?
Can you tap out a tune?
What happens if you hold a pot as you tap it?
Why does this happen?

You will need:
different-sized
  plant pots
string
nails
wooden pole
pencil

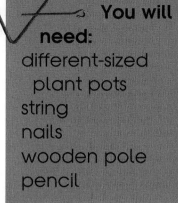

Blowing musical notes

You can make musical sounds by blowing. These children are blowing across the top of some bottles. Each bottle has a different amount of colored water in it. When the children blow across the top of a bottle, it makes a musical note. Blowing across the top of the bottle makes the air inside the bottle vibrate. A long column of air vibrates to make a low note. Short columns make high notes. Collect pictures of musical instruments that work by blowing and making air vibrate.

Make a wind instrument

You will need:
short length of plastic tubing
cork
dowel rod or stiff wire

Push the wooden dowel rod into the cork.

Push the cork into the tubing. The cork needs to fit quite tightly into the tubing. Make sure you can move it up and down with the dowel rod.

Now blow across the top of the tubing as you move the cork up and down.
What happens?
What is vibrating?
Why do the notes change?

Vibrating strings

Many musical instruments have strings stretched over a box or board. The strings are made to vibrate and make sounds. The sound box or sound board makes the sound louder. This guitarist is plucking the strings of her guitar to make sounds. Do any of your friends play string instruments? Ask them to show you how they make different notes by plucking the strings.

Make a small guitar

**You will
need:**
long cardboard
box with a lid
4 rubber bands
8 paper fasteners
scissors
pencil
small saucer
cardboard

Place the saucer on the lid
of the box and trace
around it.
Ask an adult to help cut
out a circular hole in
the lid.
Push four paper
fasteners into both ends of
the box.
Fasten the rubber bands to the
paper fasteners.
Make a cardboard "bridge" and
push it under the rubber bands.
Pluck the bands. Listen to the
sounds they make.
Stretch the bands tighter. What
differences do you notice when
you pluck the bands now?

Sending sounds

You can send sounds to your friends by speaking softly or loudly. Sometimes you need to shout to send the sound farther.
A telephone uses electricity to send sounds a long way. You can use a telephone to talk to a friend on the other side of the world!

Work with a friend
Make a simple telephone

You will need:
2 plastic cups
2 paper clips
thin string
large pin

Ask an adult to make a hole in the bottom of each plastic cup using the pin.
Thread the string through both holes.
Tie a paper clip to each end so the string does not slip through the holes.
Whisper into one cup while a friend listens through the cup at the other end.
Keep the string tight as you speak.
Can your friend hear what you are saying?
What happens if the string is loose?
How does your string telephone system work?

Bouncing sounds

Sometimes when you shout in an empty room or in a tunnel, you can hear your voice being repeated. This is the sound of your voice bouncing off the walls. It is called an echo. Find a place where you can make your voice echo.

Work with a friend
Make sound bounce

You will need:
2 cardboard tubes
stiff cardboard
ticking clock

Ask your friend to hold the piece of cardboard upright.

Place the cardboard tubes at angles to the stiff cardboard.

Put a ticking clock near one of the tubes.

Listen at the end of the other tube.

Can you hear the clock ticking?

Sound from the clock travels down one tube and bounces off the cardboard into the other tube. The ticking moves up this tube, as sound waves, to your ear.

Index

CECIL COUNTY PUBLIC LIBRARY
ELKTON, MARYLAND 21921

ELK

J
534
D

Dixon, Malcolm.
Sound and music